Dullal Ghosh
Mohammad Abu-Alqumsan
Mats Hanson

Commanding An Avatar Through Mind

AF141269

Dullal Ghosh
Mohammad Abu-Alqumsan
Mats Hanson

Commanding An Avatar Through Mind

An Approach Towards a Paradigm Shift in Adaptive Brain-Computer Interface for Robotic or Wheel-Chair Based Navigation

LAP LAMBERT Academic Publishing

Impressum / Imprint

Bibliografische Information der Deutschen Nationalbibliothek: Die Deutsche Nationalbibliothek verzeichnet diese Publikation in der Deutschen Nationalbibliografie; detaillierte bibliografische Daten sind im Internet über http://dnb.d-nb.de abrufbar.

Alle in diesem Buch genannten Marken und Produktnamen unterliegen warenzeichen-, marken- oder patentrechtlichem Schutz bzw. sind Warenzeichen oder eingetragene Warenzeichen der jeweiligen Inhaber. Die Wiedergabe von Marken, Produktnamen, Gebrauchsnamen, Handelsnamen, Warenbezeichnungen u.s.w. in diesem Werk berechtigt auch ohne besondere Kennzeichnung nicht zu der Annahme, dass solche Namen im Sinne der Warenzeichen- und Markenschutzgesetzgebung als frei zu betrachten wären und daher von jedermann benutzt werden dürften.

Bibliographic information published by the Deutsche Nationalbibliothek: The Deutsche Nationalbibliothek lists this publication in the Deutsche Nationalbibliografie; detailed bibliographic data are available in the Internet at http://dnb.d-nb.de.

Any brand names and product names mentioned in this book are subject to trademark, brand or patent protection and are trademarks or registered trademarks of their respective holders. The use of brand names, product names, common names, trade names, product descriptions etc. even without a particular marking in this works is in no way to be construed to mean that such names may be regarded as unrestricted in respect of trademark and brand protection legislation and could thus be used by anyone.

Coverbild / Cover image: www.ingimage.com

Verlag / Publisher:
LAP LAMBERT Academic Publishing
ist ein Imprint der / is a trademark of
OmniScriptum GmbH & Co. KG
Heinrich-Böcking-Str. 6-8, 66121 Saarbrücken, Deutschland / Germany
Email: info@lap-publishing.com

Herstellung: siehe letzte Seite /
Printed at: see last page
ISBN: 978-3-659-45472-1

Zugl. / Approved by: Munich / Stockholm, Technical University of Munich / KTH Royal Institute of Technology, Diss., 2012

TABLE OF CONTENTS

ACKNOWLEDGEMENT

Here we would like to thank all those who have contributed with their help, guidance, motivation and assistance in this work.

First of all, we thank LAMBERT Academic Publishing for providing us the opportunity to publish our work as a book.

This work has been carried out on behalf of KTH Royal Institute of Technology, Stockholm, Sweden at Technical University of Munich, Germany under Erasmus Exchange program and is a part of VERE (Virtual Embodiment and Robotic Re-Embodiment) European Union program.

M.Sc. Mohammad Abu-Alqumsan at TU Munich and Professor Mats Hanson at KTH Sweden have been the supervisors of this work. Thanks to Dr. Angelika Peer and Professor Jan Wikander for reviewing our work and suggesting improvements during several presentations. We are really grateful to KTH Coordinator Anna Hellberg Gustafsson for nominating this work to pursue under Erasmus exchange program.

Thanks to fellow colleagues Rigas-Georgios Zapounidis, Akhter Jamil, Abhilash Babu, Rickard Nilsson, Martin Andersson and Ertan Kayan for their company. Also to mention Pallab Rath, Shanti Swaroop Mohanty and Arjun Bhagat have motivated us for this work. Thank you Aditya Ghantasala and Áron Cserkaszky for being there with us during the work.

Last but not the least we would like to thank our parents for their love, continuous motivation and moral support...

NOMENCLATURE

Abbreviations

AUI	Adaptive User Interface
BCI	Brain-Computer Interface
BMI	Brain-Machine Interface
EEG	Electroencephalography
ERP	Event-related Potentials
ErrPs	Error-related Potentials
IErrPs	Interaction Error-related Potentials
KKT	Karush-Kuhn-Tucker
LS	Least Square
MDP	Markov Decision Process
noErrPs	no Error-related Potentials
POMDP	Partially Observable Markov Decision Process
QP	Quadratic Programming
RBF	Radial Basis Function
RL	Reinforcement Learning
ROS	Robot Operating System
SMO	Sequential Minimal Optimization
SNR	Signal-to-Noise Ratio
SSVEP	Steady State Visually Evoked Potential
SVM	Support Vector Machines

1 INTRODUCTION

In this chapter a brief introduction of brain-computer interface, brain-computer interface paradigms or techniques, problem description and book structure have been presented.

1.1 Brain-Computer Interface

People suffering from paralysis have low control over their environment. As per statistics, the number of paralyzed persons goes up to 6 million in United States alone, which means 1 in every 50 people are living with paralysis [1]. *"Movement restoration for patients with chronic stroke or other brain damage also remains a therapeutic problem and available treatments do not offer significant improvements"* [2]. The quality of life of paralyzed persons could be improved by virtue of a technique known as Brain-Computer Interface (BCI), *"often called a mind-machine interface, or sometimes called a direct neural interface or a brain–machine interface, which is a direct communication pathway between the brain and an external device"* [3]. For many decades, instrument control by just thinking, using brain waves has been accepted in science fiction. However it is only in the last few years that these systems have been shown to be feasible in laboratories [4]. An example of a person operating a robotic arm through brain waves is shown in Figure 1. In general there could be invasive or non-invasive BCIs. Invasive BCIs need electrodes to be inserted into the brain tissue [4], whereas non-invasive BCIs avoid surgery and rather use an Electroencephalographic (EEG) cap by placing electrodes on the scalp as in Figure 2. EEG measures voltage fluctuations resulting from ionic current flows within the neurons of the brain. Among several applications of non-invasive BCI, one is where paralyzed patients can spell out words and form sentences using their thoughts alone. Some other applications may include the ability of a user to control a video game by thought [5], virtual key board control [6], cursor control [7] or for neuroprosthetics applications. In the context of robotic embodiment, a paralyzed person could use BCI to navigate a robot or wheel-chair in a known environment (with known topological map), as if the robot is an avatar of the person's own body.

 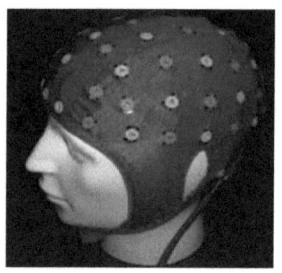

Figure 1: Example Task   Figure 2: EEG Cap (NeXus)

(FRIEND-II System–Univ. of Bremen)   (Stens Corp.–Biofeedback Equipment and Training)

1.2 Brain-Computer Interface Paradigms

Under the framework of Brain-Computer Interface (BCI), selective attention is used to extract user intentions and preferences by use of Steady State Visually Evoked Potential (SSVEP) or P300 wave. SSVEP is the oscillatory wave appearing in the occipital leads of the EEG in response to a visual stimulus modulated at a certain frequency. The frequency of the SSVEP matches with that of the stimulus and its harmonics. SSVEPs can be elicited by repetitive visual stimuli at frequencies in the range of approximately 3.5 Hz to 75 Hz [8]. In Figure 3 shown below is an example of SSVEP. In SSVEP-based BCI simultaneous flashing or flickering squares on screen are used. SSVEP has excellent signal-to-noise ratio but has limitations in the number of options that could be displayed on the screen.

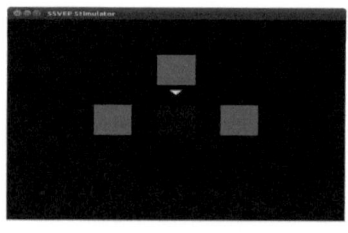 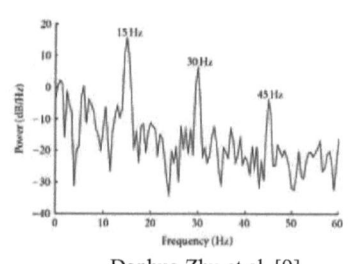

OpenViBE Software   Danhua Zhu et al. [9]

(INRIA, France)

Figure 3: SSVEP-based BCI

P300-based BCI uses Event-related Potentials (ERP) to recognize user intention. An ERP is the measured brain response that is the direct result of a specific sensory, cognitive, or motor event. More formally, it is any stereotyped electrophysiological response to a stimulus. P300 is a late appearing feature of an ERP that is elicited in the process of decision making. It is observed as a positive deflection in voltage with a latency of about 300ms [10] and hence the name P300. The signal is typically measured most strongly by the electrodes covering the parietal lobe [11]. The presence, magnitude, topography and timing of this signal are often used as metrics of cognitive function in decision making processes. It is usually elicited using a typical oddball paradigm in which low-probability target items are inter-mixed with high-probability non-target items using flashing of a matrix introduced by Farwell and Donchin [12]. Figure 4 shows the matrix at the flashing time of the items that lie in the first row from the top. P300-based BCI has higher number of options that could be displayed on the BCI screen simultaneously, compared to SSVEP-based BCI.

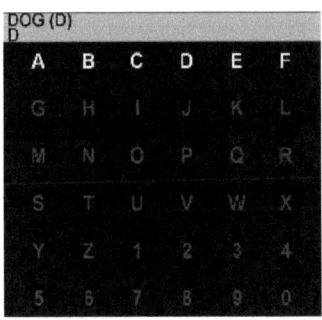

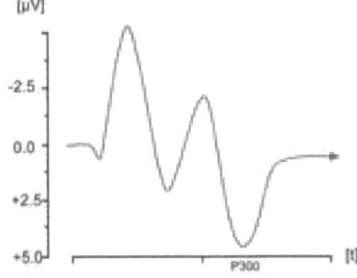

Standard 6x6 P300 speller matrix
(ETSU BCI Lab)

(Birbaumer & Schmidt, 2006, S. 481)

Figure 4: P300-based BCI

During the choice task in both of these paradigms, if the interface interprets user intention wrongly and the interface recognized option is displayed on the screen to the user, then an error-related potentials (ErrPs) is expected to be elicited in the brain. ErrPs are time-locked to commission of error or more accurately, appears after the onset of a feedback reflecting the error.

1.3 Problem Description

Semi-autonomous navigation is a flexible task sharing system where the user could assist the robot with low level navigation whenever needed. The system allows the robot to specify goals autonomously while maintaining high level path planning. The communication between user and robot is mediated by EEG signals. With the help of sensory information and a topological map of the environment, the robot is able to recognize situations when a decision should be made. Decisions are made either by the robot autonomously or by the user, depending on the level of sharing in use. In case it is recognized by the robot that a decision needs to be made by the user, the interface should provide navigational options, from which the user selects. These options could be intended goal locations or low level commands such as turn left or move forward.

BCIs, inherently have low bandwidth or bit-rate with maximum information transfer rates of 5–25 bits/min at best, which depend both on speed and accuracy [13]. Present BCI interfaces are based on fixed policy where the user, environment and the interface start always at a specific state, and based on the user input; the interface deterministically updates its contents and presents them to the user. This method of presentation to the user is often time consuming and requires much interaction load from the user and thereby diminishing the embodiment feeling of the user in commanding a robot avatar or wheel-chair in use. Therefore the first objective is to build a brain-computer interface based on adaptive policy with automatic customization feature as per the need of individual user for a navigational task, which is needed to make the best use of the scarce resources, i.e. with available options or bit-rate. The robot should be able to learn routines in user behavior and propose the navigational actions to the user in a way to optimize user interaction. The developed adaptive interface needs to be tested with the help of a Gazebo simulated environment where the robot would receive commands from the user through Robot

Operating System (ROS)[1]. The Gazebo simulator is a primary tool used for robots and robotic applications in ROS community to simulate indoor and outdoor environments. Generating a user model is also required to train and test the algorithm. The BCI system in this case is simulated by a Graphical User Interface.

In both fixed and adaptive policy based BCI system, the user intention recognition rates of classifier is subjected to error due to misclassification [14] and flows down till execution as a command unless prevented. The detection of ErrPs in the EEG signal could be exploited to prevent possible error propagation as shown in Figure 5 below. Error-related potentials raise many challenges in terms of classification due to low signal-to-noise ratio, especially in the case of single trials [15]. In Figure 6, the averages of Error-related Potential and no Error-related Potential (noErrP) signals over number of trials from 'Cz' electrode are visible to the naked eyes with clear distinction. Averaging the signal over number of trials is not a feasible method to detect ErrPs as it is generated only once within a certain duration after seeing the erroneous response of the interface. For the case of single trial, the difference between the two signals is not clear to the naked eyes as shown in Figure 7. This results in big challenge to classify and categorize the scalp-recorded EEG data for detection of ErrPs.

[1] Willow Garage, Robot Operating System (ROS)

Therefore the second objective is to recognize user's cognitive state that reflects user awareness to errors committed by the interface or, in other words, is to develop a classifier for detection of possible ErrPs in the EEG data and hence prevent misclassified user intentions or commands from being executed.

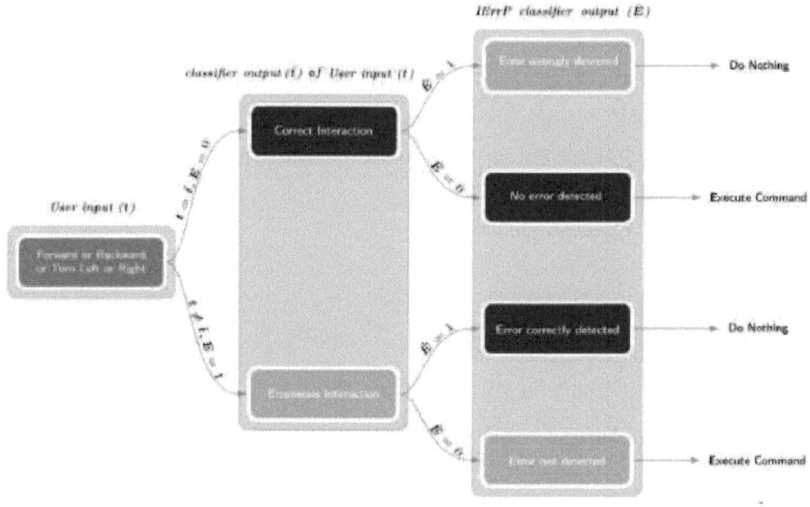

Figure 5: EEG Data Classifiers

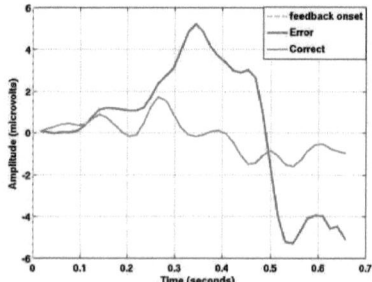

Figure 6: Average ErrP (Error trial) and noErrP (Correct trial) – [Cz electrode]

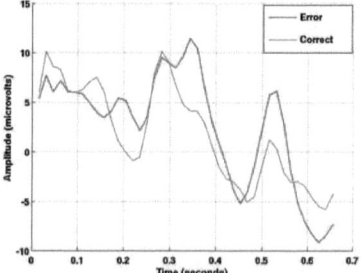

Figure 7: Single trial ErrP (Error trial) and noErrP (Correct trial) - [Cz electrode]

1.4 Book Structure

The entire book has been divided into two parts. The first part (Adaptive BCI) addresses the problem of developing an algorithm for intelligent adaptation of a BCI system as per different users. We present later in this part, the proposed algorithm and possible future works as well. The last part (Classification of Error-related Potentials), presents the state-of-the-art classifiers for detecting ErrPs in the EEG data and the methodology to use Support Vector Machines (SVM) as a classifier for detection of ErrPs. Further, the results of SVM with other state-of-the-art classifiers such as Gaussian classifier and Bayesian filter methods are compared.

2 PART ONE (ADAPTIVE BCI)

In this chapter the state-of-the-art adaptive user interface, brain-computer interface manager, the algorithms for learning and markov decision process for predicting user intention along with the training and testing results of the algorithm have been presented. At last the future works related to adaptive brain-computer interface have been described.

2.1 State-of-the-Art Adaptive User Interface

An adaptive user interface also known as AUI, is a user interface which adapts or changes, its layout and elements to the needs of the user or context. Alternatively an adaptive user interface is a software artefact that improves its ability to interact with a user by constructing a user model based on partial experience with that user [16]. An adaptive interface should have a property to only show relevant information to a user as per the context. The area of intelligent and adaptive user interfaces has been of interest to the research community for a long time [17]. *"To date, research in this field has not led to a coherent view of problems, let alone solutions"* [17]. Several studies including the work by Gajos et al. [18] which generate different interface renditions in response to different usage patterns, mention benefits of adaptive systems. Machine learning is an emerging field which could be exploited for their possible usage in designing adaptive interfaces, as has been examined in Pat Langley's work [19]. To mention, machine learning has been successfully incorporated in cell phone devices [20]. Reinforcement Learning (RL) [21], a machine learning paradigm, stands as a promising approach under the situation where the exact dynamics of the environment are not known. The learning problem is to find an optimal policy that maps states to actions, through a trial-and-error process of repeated interaction with the user. It has been successfully utilized in many applications including the problem of dialogue management [22]. RL has also been used to permit the robot to learn and optimize appropriate control policies from its interaction with the user [23, 24]. An interesting piece of work by Liu et al. [25] mentions how a mobile phone could learn context and user preferences via RL to adapt and set its alarm type automatically based on context information obtained from a variety of sensors. At another instance, the power of RL algorithm has been demonstrated [26] with application to a real-world problem as complex as controlling an autonomous helicopter. Partially Observable Markov Decision Process (POMDP) is another engineering framework that integrates Reinforcement Learning and Bayesian belief tracking and the benefits of this approach are demonstrated by the example of a simple gesture-driven interface to an iPhone application [16].

With regard to application of machine learning techniques in Brain-Machine Interface (BMI), a group of researchers have exploited RL algorithm to illustrate how BMI learns to complete a reaching task using a prosthetic arm based on the user's neuronal activity [27]. For the case of brain-computer interface system, Chavarriaga et al. have mentioned how the performance of user intention classifier could be improved with the help of RL [28]. Also POMDP model has been utilized to compute an optimal sequence of stimuli in P300 based BCI task [29]. In the case of a navigational task, it is interesting to see the research work by Perrin et al. [30] that demonstrates a method for the robot to propose low level actions to the user, like turn left or move forward, at the decision making points where the user could either accept or reject the proposition. In an unknown environment, the robotic system first extracts features so as to recognize places of interest where a human-robot interaction should take place (e.g. crossings). Based on the local topology, relevant actions are then proposed. It is to be noted that with this above methodology, only low level actions as analysed from the environment, are proposed one by one to the user and Error-related potentials in case of disapproval are used for this purpose. The dialogue management strategy continues as long as the user disapproves the propositions. Even for a single user, utilizing ErrPs to record user choice poses the problem that ErrPs classifier is often subjected to recognition error due to single trial classification and therefore multiple recordings (until the probability difference between the two most probable actions exceeds a given threshold) need to be performed to confirm the classification result. If the cardinality of goal space is large, this method seems to be

14

time consuming and involves much interaction workload to the BCI user and thus poses the risk of diminishing the embodiment feeling of the user in commanding a robot avatar or wheel-chair in use. Further, the work doesn't take care of change in user behavior and possible outlier or, in other words, it doesn't take care of uncertainties in user behavior to predict the appropriate actions and propose them to the user, which often is the case for a navigational task [31].

State-of-the-art-BCI interface is based on fixed policy which would be explained in the following section. Virtually all existing user interfaces, not only in BCI, follow the finite-state machine model [16]. The area of designing an adaptive brain computer interface with dynamic customization for navigational tasks is still not explored and is being addressed through this work. Several success stories of RL algorithms to many domains, motivates us to design an adaptive BCI system for navigational task with the help of RL, by treating the interface adaptation as an optimization problem. Further, the uncertainties in user behavior such as change in behavior and outlier could also be tackled as described in the following sections of our work. The key contribution of this work is to design a reinforcement learning algorithm to learn user behavior including possible uncertainties in behavior and predict the user intentions in a BCI based navigational task which could provide robotic embodiment feeling to the user.

2.2 Brain-Computer Interface Manager

As mentioned earlier, a paralysed person could command a robot avatar or wheel-chair using brain-computer interface system. A BCI system designed on fixed-policy interface is a finite-state automaton that mediates between the user and the robot avatar of the user. For a limited user goal-space, every goal can be mapped into one of the available BCI controls. In the case of large goal-spaces, though it is still possible to use the fixed policy approach but it is not convenient for the user due to several of the following factors. A user always starts from a starting interface and then goes down in a well-designed hierarchy which is defined by task experts till reaching the intended command, which might be a long and tedious trajectory, leading to a negative impact on the embodiment feeling. Another drawback of this approach is the requirement of a detailed task description and the fact that it imposes strong assumptions about the user and the environment by ignoring dynamics of both. Figure 8 depicts an interface designed as a finite-state automaton with three masks. Using this approach, in case the user intended option lies in the third mask, then the user needs to select 'Next' buttons in the first and second masks and in the third mask the user needs to select the intended option. So in this case, three interactions are required to decode a single user intention.

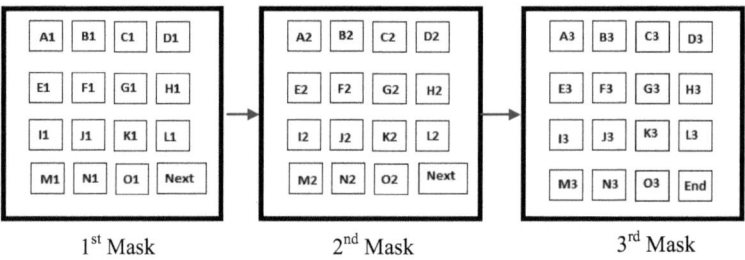

Figure 8: Example of a Finite State Automaton

2.2.1 Adaptive Interface with Dynamic Policy

As a fixed-policy interface only depends on the task description and model, the time needed to select a goal with a fixed-policy interface is user, situation and context independent. Therefore customization of the interface according to user preferences is necessary to reduce the interaction workload and consequently increase the robotic embodiment feeling of the user. This originates the necessity of designing a dynamic interface where the options proposed to the user are no longer fixed but are dynamically customized for individual user automatically. Therefore the policy based on which the interface would be designed and managed, is desired to be an optimized policy. This could be formulated as an optimization problem where decision needs to be made by the interface manager to propose the optimized set of options to the user at every decision making situation.

2.3 Markov Decision Process and Reinforcement Learning

Although, human has the potential for relatively random patterns of behavior, there are easily identifiable routines in every person's life [31]. Based on this, the proposition of best set of actions to the user could be modelled as a Markov Decision Process (MDP). MDP provides a mathematical framework for modelling decision making in situations where outcomes are partly random and partly under the control of a decision maker. More precisely MDP is *"a decision-theoretic model capable of taking into account both uncertainty in the effects of its actions and trade-offs between competing short-term and long-term objectives when making decisions"* [32]. Here, S is the set of all states, A is the set of all possible actions and P is the set of all state to action selection probabilities. At each time step, the process is in a state $\in S_t$, and the decision maker may choose any action $a \in A_t$, that is available in state s with probability $p \in P_t$ from that state. The process responds at the next time step by randomly moving into a new state $s' \in S$, and giving the decision maker a reward R. The next state s' depends on the current state s and the decision maker's action a. But given s and a, it is conditionally independent of all previous states and actions for first order Markov model or, in other words, the state transitions of a MDP possess the Markov property. MDP is modelled with $\{S, A, P, R, \gamma\}$, where γ is the future discount factor for rewards, with a value typically close to 1. In case the dynamics of the environment or state transition probabilities are not known in prior, the system should learn these transitions explicitly by interacting with the environment. Therefore this learning mechanism which gains knowledge of environment by virtue of interacting with its environment falls under the category of Reinforcement Learning (RL) scheme. The learning of the algorithm should lead to a policy that would map a state to an action for all possible states.

In our present work a P300 interface based on fixed policy has been implemented that presents, in a predefined order, to the BCI user a sequence of command option sets to choose from the same. It is worth mentioning that the total number of possible command options exceeds the maximum number of goal locations that can be hosted by a P300-based BCI mask. The user is modelled as a MDP and the interface should learn the user behavior or actions carried out from every state and propose a set of best possible actions to the user as per the capacity of P300-based BCI mask, in other words, it should capture the user intentions. Here action $a \in A$ is the proposition of an action to the user. It is to be noted that actions here are propositions with respect to the interface manager, i.e. what set of commands should be presented to the user at each decision making situations. The scenario used here is a pure navigation scenario, where all user goal-oriented commands are of the form "move to location x" or low level commands such as turn "left" or move "forward" as shown is Figure 9 for few state and action pairs.

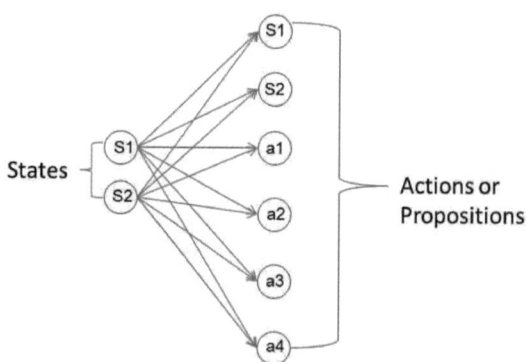

Figure 9: State-action Mapping

Additionally, RL-based algorithm has been developed thereby aiming at finding an optimal policy for the interface adapted as per individual user. The results of both approaches have been compared in terms of user interaction workload which is measured by the time needed to get user intention decoded by the system. In both

18

cases, i.e. adaptive and fixed-policy interface, however, user input is mapped into user commands that trigger the transition to new states and interface actions without considering classification errors. For tasks with available detailed description and model or where the adaptive policy assigns same priorities to multiple actions, fixed policy will be used for those tasks or actions. This is because it is intuitive to go for fixed policy if no concrete decision could be made regarding the possible user intended actions. This results in a system with mixed policy considering both fixed and adaptive policies. Therefore the interface proposes actions to the user as per the mixed policy. Figure 10 relates the adaptive interface manager to other system components such as EEG decoder to recognize user intention I_u and present the decoded intention to the user by the help of interface manager. The noise N_i is due to the disturbances in the EEG signal as it is captured from the electrodes placed at the scalp of the user and also includes the user intention classification error or noise by the classifier. The recognised user intention in P300-based or SSVEP-based BCI, is sent as a robotic command to the robot avatar or wheel-chair in use.

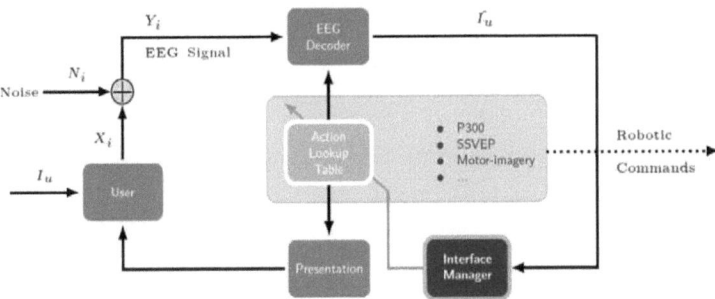

Figure 10: Adaptive Interface Manager and its
Relation to Other Components of the Embodiment

2.3.1 Learning User Behavior and Prediction of User Intention

Decision-making model is used here to first observe and learn the user behavior and then predict the user intention in the context of a navigational task. The underlying pattern in user behavior could be of first, second or higher orders, though no assumption has been made with respect to the order of the dynamics. Modeling user behavior with second or higher order Markov Decision Process, poses the problem of increased memory consumption. Therefore a first order MDP is considered to solve the navigation problem. The knowledge of the user model is incomplete which means that the action selection probabilities from a state are not known in advance and they need to be explicitly learned from the user behavior using the strategy of reinforcement learning.

States are defined as every possible goal locations in the map. Actions are propositions of next states or low level commands to the user. These low level commands include commands like moving forward or turning left etc. Thus the actions are more in numbers than the states. Both the number of states and the number of possible actions are known in prior. The number of state-action pairs thus rises to number of states times' number of actions. The objective is to find a policy which could map each state to a best possible action. As P300 mask could contain not just one but a set of possible actions or propositions, therefore for each state the policy should rank all user intended actions and show a set of most probable actions in the first mask depending upon the capacity of the designed P300 mask which is available in different variants.

2.3.2 Learning Behavior

The learning mechanism associated with navigational task is a continuous process having finitely large goal state space $S \in \{s1, s2, s3, ..., sn\}$ of the robot and interface action space $A \in \{a1, a2, a3, ..., an\}$. Every goal state is considered as a location in the map which is fully observable. The user action is fully observable but the action probability P from a state is not known in prior. The algorithm has the opportunity to learn regarding user behavior after every interaction with the user. It is interesting to note, as every state here is a goal location and therefore the problem of predicting with finitely large goal spaces could be divided into sub-problems each consisting of only two goal locations or states, one as a starting and the other as ending state. In other words, each sub-problem could be considered as an episodic task where the episode ends after every transition. Based on the interface action, the interface receives an immediate reward r, which is considered here as the needed time or number of interactions with the user. Though there are several possible actions from a state, the user only performs one action every time and thus reaching from a starting state to a goal state in one step. The user behavior is stochastic with actions having a probability distribution from each state. For the purpose of learning a stochastic user model, sampling is done from the probability distribution of actions in a state and a trajectory generated is made available to the algorithm. The algorithm is expected to learn the user model from this generated trajectory. Because the sampled trajectory is made available to the algorithm, bootstrapping method is not used in this case, where a guess is made depending upon another guess as in the case of Temporal Difference prediction method [21]. It is to be observed that here every transition in the trajectory involves a learning step for the algorithm.

The state of the robot (location) at time t is defined as $s \in S_t$. Taking action $a \in A$ from state $s \in S$ will either make the robot move to a state s' if $a \in A_h$, where A_h is the set of all high level commands or, in other words, the set of all goal locations or will make no change to s if $a \in A_l$, where A_l is the set of all low level action commands. More formally, S_t is defined to be a random variable referring to current state and S_{t+1} refers to next state. A_t is a random variable that refers to an action at time t and hence;

$$S_{t+1} = S_t \text{ if } a \in A_l$$
$$S_{t+1} \neq S_t \text{ if } a \in A_h \text{ or } a \in S_{t+1} \tag{1}$$

We assume that that robot executes all commands deterministically, which means that if we know a then we know s' deterministically. Therefore the problem of predicting the user intended next state is exactly the same as predicting the next user action. Based on this transition, the state-action pair receives a positive reward of $r_f = 1$. A term defined as frequency of selecting an action a from a state s is the accumulated rewards over time and updated for the occurred state-action pair as follows:

$$F(a/s) = F(a/s) + r_f \tag{2}$$

As the robot has no initial knowledge regarding the dynamics of the environment, the value of each state-action pair is termed as:

$$Q(s,a) = 0 \ \ where \ \forall s \ \in S \ and \ \forall a \in A \tag{3}$$

It is worth mentioning that frequency of a state-action pair $F(a/s)$ represents the $Q(s,a)$ value of that state-action pair and therefore the below assignment holds true for this case here:

$$Q(s,a) = F(a/s) \tag{4}$$

A policy is defined as a mapping from every state s to an action a. Policy improvement is done by making the policy greedy with respect to the current value function. For any state-action value function Q, the corresponding greedy policy is the one that, for each $s \in S$, deterministically chooses an action with maximal Q value:

$$\pi(s) = argmax_a Q(s,a) \tag{5}$$

The reward r is a form of motivation for the occurred state-action pair and is likely to increase the chances of occurrence of the action a while the robot visits the state s next time. As an example, after first interaction with the user, the Q value of the occurred state-action pair at time $= 1$, is updated to:

$$Q(s,a) = 1 \tag{6}$$

22

Matrix of dimension equal to the number of states times number of actions, is memorized which contains the $Q(s, a)$ values $\forall s \in S$ and $\forall a \in A$, and the values are updated after every interaction with the user.

2.3.3 Change in Behavior

Human has potential for relatively random patterns of behavior and hence the behavior could be termed as stochastic [31]. The amount of randomness regarding behavior of a user corresponds to its entropy. Entropy (in bits) of a discrete random variable X is defined by Claude Shannon in the equation below.

$$H(X) = - \sum_{i=1}^{n} p(x_i) log_2 p(x_i) \tag{7}$$

People who live less entropic lives, it is easier to predict their behavior but who live high entropic lives tend to be more variable and harder to predict. One such case of user behavior could be related to change in behavior after certain number of interactions with the interface where the user jumps from a policy π to another policy π' and continues following policy π' thereafter. In order to handle this situation of changed behavior by increasing the convergence rate of the algorithm or learn the altered behavior faster, another form of reward is incorporated. Separate interaction counter or local timer for each state is defined and an action executed in a state increments the timer associated with that state by one. This local timer value is rewarded to the occurred action of the corresponding state. Therefore this timer associated reward function $T(s, a)$ for a state-action pair is intuitively a function of the time the action a happened last in the corresponding state s. It is worth noting that the timer associated reward function $T(s, a)$ is equal in magnitude to the marginal count $F_{t_a}(s)$ of the state s at the time of last occurrence of the action a in the same state s. Hence each state-action pair is time stamped with a reward value equal to the marginal count of the corresponding state when the respective action is seen last occurring in that same state and hence defined mathematically as below:

$$T(s,a) = F_{t_a}(s) \quad where \; \forall s \in S \text{ and } \forall a \in A \tag{8}$$

Therefore the modified $Q(s,a)$, when the possibility of changed behavior pattern of a user is included, takes the form:

$$Q(s,a) = w_f * F(a/s) + w_t * F_{t_a}(s) \tag{9}$$

where w_f is the weight of Frequency based reward, w_t is the weight of Timer based reward

While the first term on the right hand side of the above equation favors frequent actions as it is related to frequency of taking an action in a state, the second term favors recent actions in a state. The intuition behind the second term on the right hand side of the equation is that in a state the action that occurred recently is assigned higher timer value than the action occurred earlier in the same state. Therefore the recent action or user behavior is given higher priority compared to the previous action from the same state.

It could be noted that the use of a global timer or interaction counter as a reward function is avoided here, though it seems easier with regard to implementation. It is due to the fact that using a global timer would assign a large value to a recent action that occurred from a less visited state in which case the reward associated with the global timer would be more dominant than the reward associated with the frequency term in spite of having the weighting terms for each of timer and frequency. Hence for the case of a global timer, finding weights that would compensate for both less visited as well as frequent states would be difficult and may lead to wrong prediction of user intentions. If there is no change in user behavior observed then also a global timer based reward function would perform poor in terms of prediction, compared to a local state timer based reward.

2.3.4 Outlier in Behavior

Outlier is defined as completely random pattern in user behavior or action for certain short period of interactions such as visiting a 'Friend' once in a month etc. In other words, the user follows a policy π and after certain number of interactions the user started following or switched to another policy π', having completely random state-transition probabilities for few number of interactions and then again goes back to the original policy π. The difference between change in behavior and outlier is that, in the case of first one the user never goes back to the original policy π after changing the behavior to another policy π', while for the latter case the user goes back to the original policy π after certain number of interactions. The weight of timer w_t introduced in equation (9) above helps the algorithm to learn outlier as it assigns higher timer associated reward to recent transition but in order to increase the rate of learning during outlier period and also after completion of it, another reward function is defined which assigns a reward $r_i \in \{0,1,2\}$, based on the number of interactions needed to select user intended action a available in a state s. Here the number of interactions means the number of masks that are needed to be shown to the user for selection of an action from a given state. Hence r_i is a measure of the immediate reward from user to the interface policy. The cumulative interaction reward function for a state-action pair $I(s,a)$ is memorized in a table, which is simply the accumulated immediate interaction rewards for an action a that occurs in state s and is updated after every interaction with the user as shown below:

$$I(s,a) \leftarrow I(s,a) + r_i, \ where \ \forall s \in S \ and \ a \in A \qquad (10)$$

In our experimental setup if the user intended option appears in the first mask then immediate interaction reward r_i for that state-action pair is assigned a value 0 and similarly if it happens to be in second or third mask, it is rewarded with values 1 and 2 respectively. The intuition behind introduction of this interaction based reward function is to allow the algorithm to converge faster, mainly after outlier period is over, by motivating the user intended actions that need greater number of masks to be shown to the user.

Combining above described three kinds of situations, i.e. considering a stochastic user behavior with change in pattern as well as presence of outlier in behavior, the $Q(s, a)$ could be reformulated as a combination of rewards based on state-action frequency, local state timer and number of needed interactions, as follows:

$$Q(s, a) = w_f * F(a/s) + w_t * F_{t_a}(s) + w_r * I(s, a) \tag{11}$$

where w_f is the weight of Frequency based reward, w_t is the weight of Timer based reward and w_r is the weight of needed Interaction based reward

2.3.5 Prediction of User Intention

Decision needs to be made as per the optimal policy of interface that imitates the user hidden policy. The action based on optimal policy π^*, is searched in the look-up table of $Q(s, a)$. As in the case of P300-based BCI, there is possibility of displaying a set of actions in the mask, therefore the optimal policy π^* in this case ranks all the actions, with higher Q valued actions at the top for every state and proposes the actions to the user as per the capacity of the mask. Mathematically:

$$\pi^*(s) = \arg max_{set \ a \in A} Q(s, a) \tag{12}$$

As the actions are ranked as per the relative values of $Q(s, a)$, therefore the expression:

$$Q(s, a) = w_f * F(a/s) + w_t * F_{t_a}(s) + w_r * I(s, a),$$

could be transformed into $Q'(s, a) = F(a/s) + w'_t * F_{t_a}(s) + w'_r * I(s, a) \tag{13}$

In the following sections, $Q'(s, a)$ is mentioned as $Q(s, a)$ for simplification. The process of learning and prediction steps could be viewed together as shown in Figure 11 and the algorithm is also mentioned below.

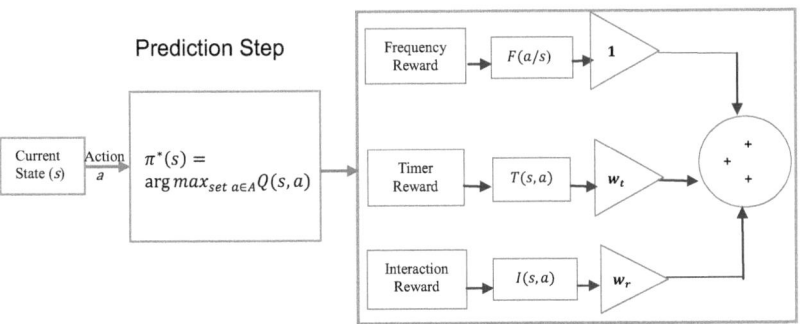

Figure 11: Process of Learning and Prediction

$begin$

$Initialize\ F(a/s),\ F_{t_a}(s),\ I(s,a) \leftarrow 0\ \ \forall\ s,a$

$repeat\ for\ each\ iteration$

$Initialize\ r \leftarrow 0,$

$\qquad\qquad A \leftarrow set\ of\ all\ actions \qquad\qquad\qquad // r\ is\ immediate\ interaction\ reward$

$\qquad\qquad Q(s,a) = F(a/s) + w_t * F_{t_a}(s) + w_r * I(s,a) \quad // prediction\ step$

$\qquad for\ i = 1:3$

$\qquad\qquad A_l = \pi^*(s) = \quad arg\ max_{A_l \in A}\ \sum_{a \in A_l} Q(s,a)\ \ s.t.\ \ A_l\ contains\ non-$

$\qquad\qquad\qquad\qquad\qquad\qquad\qquad\qquad repeated \qquad elements \qquad and \qquad |A| =$

$P300\ mask\ capacity$

$\qquad\qquad if\ a_u \subset A_l \qquad\qquad\qquad\qquad // user\ intended\ action$

$\qquad\qquad\qquad break$

$\qquad\qquad else\ if\ a_u \leftarrow next\ (IErrP = 1)$

$\qquad\qquad\qquad increment\ r\ by\ 1$

$\qquad\qquad\qquad A \leftarrow A - A_l$

$\qquad\qquad end\ if\ loop$

$\qquad end\ for\ loop$

$\qquad\qquad F(a/s) \leftarrow F(a/s) + 1$

$\qquad\qquad update\ F_{t_a}(s)$

$\qquad\qquad I(s,a) \leftarrow I(s,a) + r$

$\qquad continue$

end

Table 1: Learning and Prediction Algorithm

2.4 Training of the Algorithm Parameters and Testing

A map of the environment is built in Gazebo with several goal locations. Each of the location like 'Office', 'Hostel' etc., are assumed to be goal locations or states of the PR2 willow garage robot, in this case. In total, thirty nine states are defined in the map. Actions are thirty nine high level goal based and four low level commands with total of forty three actions to be chosen among. Robot Operating System is used to publish the desired goal location as a ROS topic to PR2 navigational stack or to command the robot through low level navigation. The path planning algorithm is made available with PR2 navigational stack. A top-view snap shot of the simulated Gazebo world with PR2 robot is shown below in Figure 12.

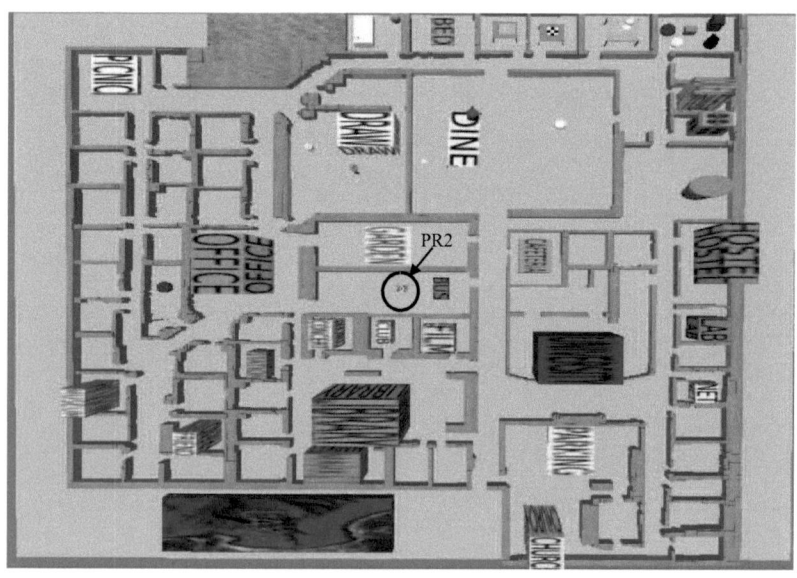

Figure 12: Simulated Robot World in Gazebo

As described earlier the user model includes stochastic behavior. The probability of each action in a state are defined for the purpose of developing a user model. The user is modeled as first order MDP and implemented as shown in Appendix A. Three different user models are developed in order to test the performances of the algorithm with different kinds of models. The first model which is a high entropic model, has uniform distribution with average of maximum probabilities around 0.04 while in the second user model with medium entropy, it is assigned a maximum state-action probability as 0.5 on an average. The last user model has 80% deterministic behavior and thus lower entropic model, to prefer a particular action from a state. In order to introduce changed behavior mode, the user model has the provision to define number of training instances after which the behavior pattern is changed. Two modes for changed behavior pattern are considered. During strong change in behavior mode, all the state-action probabilities are redefined with high entropy while during weak change in behavior mode; only the action with maximum probability from each state is swapped with another action available in the same state. In the user model, there is also provision of defining outlier period during which the user follows a completely random policy. All together, the three user models are equipped with different maximum state-action probability distributions and simulate change in behavior and outlier. For fixed policy interface, the user model is used to evaluate the testing performance while for adaptive policy it is used both for training and testing.

2.4.1 Fixed Policy Interface

For our test case, we have defined an environment with 39 states and 4 low level commands and this sum up to 43 actions in total to simulate a P300-based BCI with 3 masks. Each mask could contain at most 16 options including options such as 'Next', 'Back' or 'Repeat' buttons. As it is expected, there is no 'Back' button introduced for the first mask which means it has 15 possible actions including the initial robot pose which is displayed as 'Robo_initial_Pose'. The 'Repeat_Options' is dedicated only for the mask that comes in the last. In the case of fixed policy interface, the BCI masks have predefined options in a specified sequence. We have used the map of the environment to define a fixed policy. The masks designed in QT platform used for fixed policy P300-based BCI are shown in Figure 13 below.

1st Mask

2nd Mask

3rd Mask

Figure 13: (a), (b) and (c) Simulated BCI with 3 Masks
Used for Testing of Fixed Policy P300-based BCI

2.4.2 Adaptive Policy Interface

In order to train the algorithm, the trajectory developed with the help of a user model as described earlier, is used. The transitions in the trajectory are used as training instances, sequentially starting from an initial state till the length of the trajectory. After each instance of training, the algorithm is tested against the number of interactions needed in order to capture the immediate next action while following the trajectory.

2.4.3 Results of Fixed and Adaptive Policy Interfaces

For the user model without incorporating change in behavior and outlier, result could be seen below in Figure 14.

(Note: The abscissas of the following plots have different ranges)

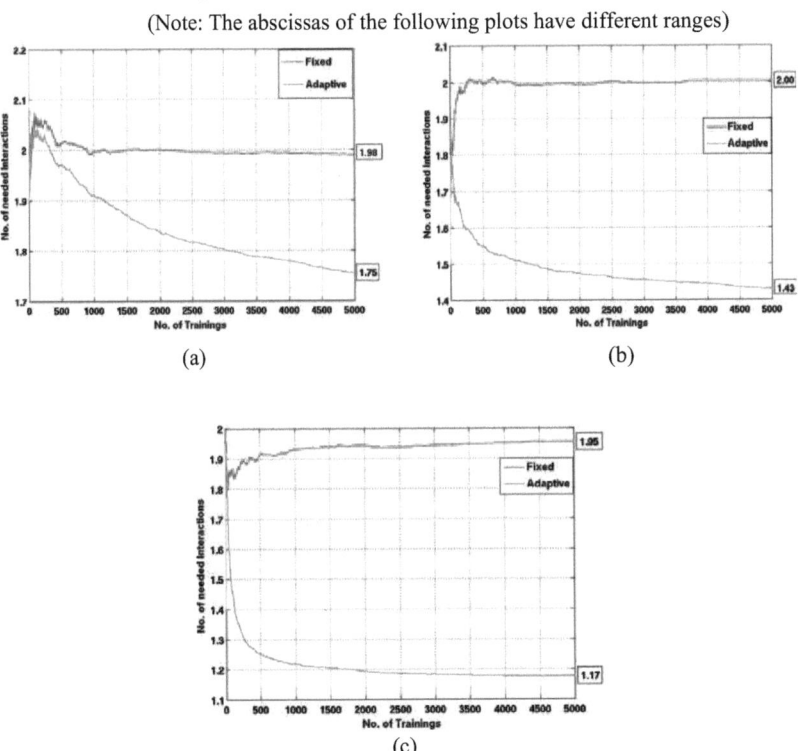

(a)

(b)

(c)

Figure 14: User Behavior: (a) Highly Entropic Behavior, (b)
Medium Entropic Behavior and (c) Low Entropic Behavior

31

For the fixed policy case, the average number of needed interactions is higher than the interactions needed for adaptive policy in all of the three kinds of user models described above.

It could be seen for the model with higher entropy which is a measure of randomness, the learning rate is slow and it approaches a higher steady state value after certain number of trainings, while in the case of medium entropic behavior it attains a lower steady state value. As expected, in that third case with low entropic behavior, the steady state value reaches near to the ideal interface which requires only one interaction to capture user intention. So from the results above, it could be stated that a higher entropy in user behavior leads to a lower convergence rate of the learning algorithm or higher average time per interaction. It means that if there is no routine in user behavior, there is no learning of the algorithm.

Different values of timer weights have been analyzed and the results are displayed both for weak and strong change in behavior separately in Figure 15.

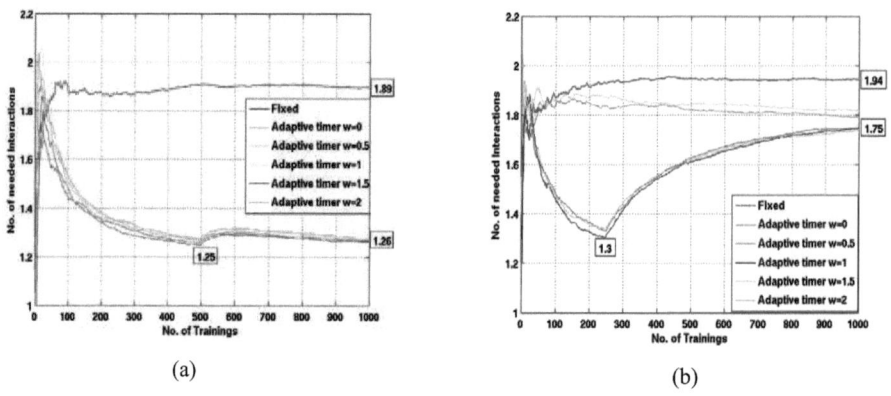

Figure 15: Changed Behavior Mode: (a) Weak Change at 500 number of trainings, (b) Strong Change at 250 number of trainings

As is in the previous case, fixed policy performs poorer than the adaptive policy even in the case of change in behavior. It is worth noticing that in the case of weak change in behavior, the algorithm quickly converges even after the changed behavior pattern is introduced and approaches a value close to the steady state value observed before there is change in behavior. But for a strong behavior change, the learning curve approaches towards the case of completely random behavior as shown in Figure 14 (a). Different timer weights affect the learning rate which could be seen in Figure 15, above.

For the situation where there are outliers in user behavior pattern, the results are as follows in Figure 16.

(Note: The abscissas of the following plots have different ranges)

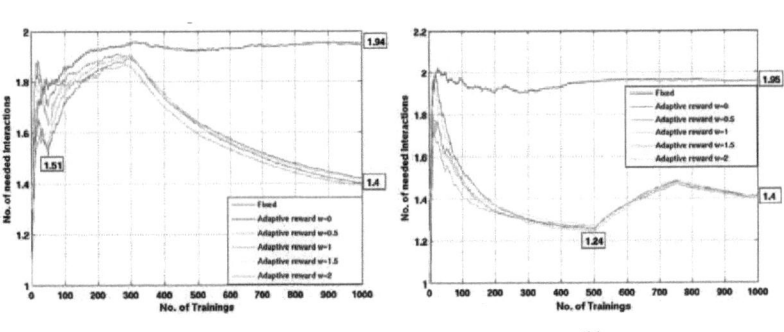

(a)   (b)

Figure 16: User Behavior with Outliers (a) from 50 to 300 number
of trainings and (b) from 500 to 750 number of trainings

As expected, there is loss of information for the cases of early occurrence of outlier and also for later occurrence of outlier in user behavior. After the outlier period is over, learning curves in both cases approach towards the same value. Again, different weights for cumulative interaction reward contribute to different convergence rates and steady state values. An attempt is made to analyze the performance of the algorithm where there are both change in user behavior and outlier and the results for two completely different user models, are as follows:

33

Outlier (50-300 number of trainings) and weak behavior change at 500 number of trainings

(a)

Outlier (250-750 number of trainings) and weak behavior change at 900 number of trainings

(b)

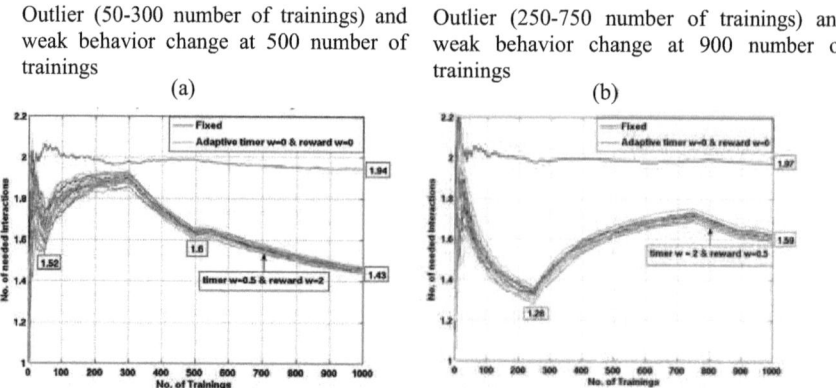

Figure 17: Both Outlier and Change in Behavior: (a) and (b)

Even with these combination of outlier and change in behavior cases, fixed policy is observed to be performing poorer compared to adaptive policies evaluated based on different combinations of weight for timer and interaction reward. It could be stated that if the outlier period is longer, the loss of information is more. However different weights could be set with regard to timer and interaction reward, which affect the convergence rate and steady state value or steady state error. It is worth noting that setting positive values for w_t and w_r helps the algorithm to converge faster both during changed behavior mode and also during outlier period but they need to be tuned individually in order to lower the steady state error. Therefore designing individual weights is a trade-off between convergence rate and steady state error of the learning algorithm. For particular user behavior, these weights are set manually after observing the steady state value of the algorithm for different set of weights. The algorithm could further be enhanced by automatic adaptation of the weights for different users. In order to provide this enhanced capability, number of iterations need to be carried out and the weight values need to be selected which satisfy the convergence criteria, i.e. to select those weight values which help the algorithm to converge within specified number of iterations and within a defined steady state value.

The training and testing method mentioned above could be viewed as an on policy training and testing, because prediction is done after every interaction with the user model. However, testing is also carried out after the learning period is over or the algorithm is exposed to a specified number of instances. Thereafter, it is tested for a trajectory of length twenty and the observed average number of interactions closely matched the interaction values seen from the above figures. Adaptive policy again performed better with a needed average interaction value of 24 compared to the fixed policy which is 38 on average for lower entropic user behavior of Figure 14 (c) above.

2.4.4 Extension of the Learning Scheme

The first order markov decision model is extended to second order markov decision model, as it is believed that implementing a higher order markov model may better predict the user behavior which could be of any order in reality. However a higher order markov model increases the memory consumption rapidly. As an example, in a second order markov model, the next state of the user not only depends on the present state but also on the immediate state prior to the present state. It consumes memory of approximately $(|S|^3)$, where $|S|$ is the cardinality of the state space. Therefore due to increased memory consumption and also due to unavailability of second order MDP of user model, this learning algorithm has not been tested. There is an attempt made to extend the definition of state to include time schedule, in order to have better prediction of user behavior. This is because user behavior is dependent not only on current state but also on time of the day, e.g. a finite number of temporal values, twenty four in our case considering an entire day with twenty four hours, is used to discretize each state which is originally a goal location into finite number of states to include both location and time of the day in the definition of a state. Separate track of locations along with hours of the day are maintained in the memory. However testing of the algorithm with time schedule would need a more complex user model than implemented here in our case and increased memory usage due to state discretization.

2.5 Future Work on Adaptive Brain-Computer Interface

As part of the future work, this algorithm could also be enhanced with the ability to distinguish a goal location to a sub-goal location where a sub-goal might be triggered due to external events such as ringing of a bell, phone etc. This could be implemented by ignoring the look-up table Q value update rule when a sub-goal is recognized as an external event. The Markov Decision Model could be extended to include partial observability of user state which might arise due to misclassification of user intention in the BCI task. Another improvement in the algorithm might be related to current memory requirement of the algorithm as it is evident that the memory requirement increases rapidly depending upon the number of states. This is because of the fact that the look-up table memorizes values for each state-action pair. In order to reduce the memory requirement, sparse representation might be implemented. Also the definition of state could further be enhanced by including additional sensory information that might be input to the learning algorithm such as environmental temperature condition, light intensity etc. in order to better predict user behavior, as these parameters also affect user behavior and hence user intentions. But it would require additional memory for implementation. Further to enhance the optimality of this BCI, the reward in terms of cognitive load or embodiment feeling of the user could also be incorporated in the learning algorithm.

3 PART TWO (CLASSIFICATION OF ERROR-RELATED POTENTIALS)

In this chapter state-of-the-art classification of error-related potentials, introduction to support vector machines, feature extraction and classification method using support vector machines, training data set used for classifier, and the performance results have been presented. This chapter ends with describing the potential future work related to support vector machines classifier for detection of error-related potentials in electroencephalographic data.

3.1 State-of-the-Art Classification of Error-related Potentials

Interaction Error-related potentials (IErrPs) are special features that can be detected in the EEG, after a wrong action selection by the BCI system or the user. After the onset of the feedback indicating the selected action, these features can be distinguished by first, *"a sharp negative peak after 250ms followed by a positive peak after 320ms and a second broader negative peak after 450ms"* [33]. Figure 18 and 19 show both ErrP and noErrP signals for average and single trail cases respectively. In order to detect the presence of ErrPs in the EEG data, a classifier needs to be developed.

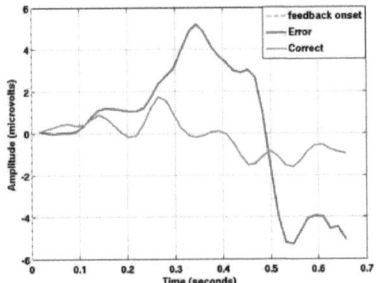

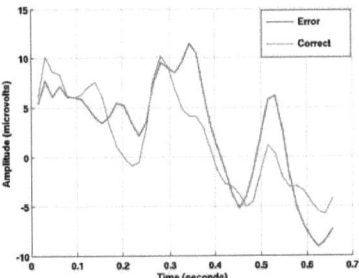

Figure 18: Average ErrP (Error trial) and noErrP (Correct trial) – [Cz electrode]

Figure 19: Single trial ErrP (Error trial) and noErrP (Correct trial) - [Cz electrode]

Out of several methods of data classification, as there is no optimal generic method, the method should be chosen depending on the task. If the training data is labelled, then supervised learning methods are used for classification and if the training data is unlabelled then it falls under unsupervised learning category which tries to find hidden pattern in data. A supervised learning algorithm analyses the training data and produces an inferred function, which is called a classifier. From training samples a decision rule can be constructed that accurately assigns labels for feature vectors in new sets of data, e.g. Support vector machines (SVM) or Gaussian classifier. Literature survey revealed that that previous work on EEG data classification for detection of ErrPs carried out by Bollon et al. [15] is able to recognize 71% of ErrP and 85% for noErrP data using a Bayesian filter. The study by Chavarriage et al. [28] has reported the maximum recognition rate of Gaussian classifier for three different subjects around 92% and 73% for true positive and true negative respectively. Similarly the average of the results obtained from three different subjects, reported by Ferrez et al. [34] are around 79% and 82% with standard deviation of 6.6 and 7.0 percentages for true negative and true positive respectively, as shown below in Table 2. In their statistical classifier, every Gaussian unit represents a prototype of one of the classes to be recognized.

Chavarriage et al.			Ferrez et al.		
Subject	ErrPs (%)	noErrPs (%)	Subject	ErrPs (%)	noErrPs (%)
1	73.5	92.01	1	87.3 ± 11.3	82.8 ± 7.2
2	58.91	83.82	2	74.4 ± 12.4	75.3 ± 10.0
3	66.29	86.86	3	78.1 ± 14.8	89.2 ± 4.9
Avg	66.23	87.56	Avg	79.9 ± 6.6	82.4 ± 7.0

Table 2: State-of-the-art results of Error related Potentials classification

3.2 Support Vector Machines

Support Vector Machines is a method that uses supervised learning for analysing and recognizing patterns in data both for classification and regression. SVM emerged in mid-1990 from the area of statistical learning theory developed by Vapnik in the late 1970's [35]. SVM has several benefits compared to other classification techniques [36]. Today SVM are widely used in many areas, for handwritten digit recognition, object recognition and many others. *"Support Vector Machines are among the best (and many believe, it is indeed the best) "off-the-shelf" supervised learning algorithm"* [37] and therefore among various available methods of classification, SVM is chosen to classify EEG signal for detection of error-related potentials elicited in the brain after observing erroneous response. SVM is a representation of the examples as points in space, mapped so that the examples of the separate categories are divided by a clear gap that is as wide as possible as shown in Figure 20. It constructs a hyperplane or set of hyperplanes in a high or infinite-dimensional space, which is used for classification. Intuitively, a good separation is achieved by the hyperplane that has the largest distance to the nearest training data point of any class, also called as functional margin; since in general the larger the margin the lower the generalization error of the classifier. A functional margin is defined as follows:

$$y^i(w^t x^i + b) \tag{14}$$

where x^i is the training instance, y^i is the corresponding class, w^t is the weight vector with b as the intercept term.

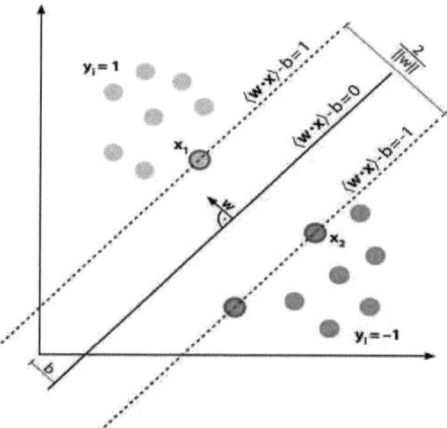

Figure 20: Support Vector Machines Example [38]

New examples are then mapped into that same space and predicted to belong to a category based on which side of the gap they fall on. For linearly separable data, a Linear SVM classifier works well. In addition to performing linear classification, SVMs can efficiently perform non-linear classification using what is called the kernel trick, implicitly mapping their inputs into high-dimensional feature spaces. Whereas the original problem may be stated in a finite dimensional space, it often happens that the sets to discriminate are not linearly separable in that space. For this reason, it is proposed that the original finite-dimensional space be mapped into a much higher-dimensional space, presumably making the separation easier in that space. To keep the computational load reasonable, the mappings used by SVM schemes are designed to ensure that dot products may be computed easily in terms of the variables in the original space, by defining them in terms of a kernel function $K(x, y)$ selected to suit the problem. Therefore selection of kernel function is an important step in the process of classification. This problem of kernel and kernel parameter selection could be simplified because simple kernels have proved to be sufficient enough and appropriate parameters can be found using grid or pattern search. Among various, one of the advantages of SVM is that it could find a boundary between the classes even for data which are not linearly separable, by projecting them into higher

40

dimensions. Also SVM provide a good out-of-sample generalization, if the kernel parameters are appropriately chosen. For the sake of these above mentioned advantages, SVM classifier is chosen for classification of EEG data to detect presence of error-related potentials.

3.3 Data Set Used for Classifier

In this simulated BCI experiment similar to the one described in [34], it has been attempted to explore interaction ErrPs in case of erroneous keyboard interactions. Thereby the user tries to push a ball into a hole which is located on the same horizontal line as the ball using keyboard left and right arrow keys only. The user input is translated by the interface into movements of the ball; thereby it moves the ball into the wrong direction with a probability of error P as shown in Figure 21. The recognition of the ErrPs is challenging due to the low signal-to-noise ratio (SNR) inherent in single trials, as opposed to averaging number of trials in the case of P300.

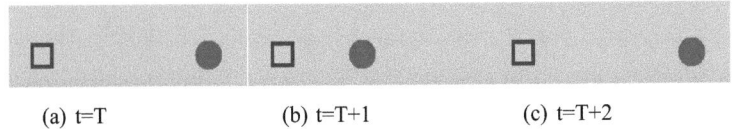

(a) t=T (b) t=T+1 (c) t=T+2

Figure 21: Experimental setup for interaction ErrPs. Using left and right arrow keys, the user should bring the ball (red circle) into the hole (blue rectangle). As an example, the initial positions of the ball and the hole, are shown in (a). The user hits the left arrow button to move the ball closer to the hole, and the result is shown in (b). Interaction ErrPs are evoked when the user hits the left arrow button and the ball goes to the right as shown in (c). Erroneous interactions reduce the information transfer rate ITR, e.g. ITR=0 for this example.

Ignoring user interactions, which are followed by IErrPs, leads to higher information transfer rates [33, 34, 39]. As a pilot study, data is recorded from one subject using a 32-channel acquisition system (from g.tec company). Based on the extended 10-20 system, 32 active electrodes are placed at the following positions: FP1, FP2, F7, F8, F3, F4, T7, T8, C3, C4, P7, P8, P3, P4, O1, O2, AF3, AF4, FC5, FC6, FC1, FC2, CP5, CP6, CP1, CP2, Fz, FCz, Cz, CP, Pz, Oz. One additional passive ground electrode is placed at Fpz. Signals are referenced to the right earlobe. The signal is sampled with frequency of 256 Hz. EEG data is often subjected to noise which may arise due to blinking of eyes or movement of body parts. EEG data therefore need to undergo pre-processing, before it could be used for classification. Pre-processing includes operations like artefact removal which is a noise with high magnitude. Then the signal is re-sampled which applies an anti-aliasing filter or low pass finite impulse response filter to the signal and changes the sampling rate to one fourth of original sampling value, i.e. to a sampling frequency of 64 Hz. Among several available channels, data from channel 'Cz' which is located near the parietal lobe that has strong signal content is used to train the classifier. Over 5 sessions total of 74 and 260 instances of ErrPs and noErrPs are captured respectively for training, validation and testing of classifier. A parameter 'time of interest' is used to extract each signal portion of interest which is set between 0 and 650ms here. Implementation of SVM is carried out the following way.

3.4 Feature Extraction and Training of Classifier

3.4.1 Method of Feature Extraction

It is one of the most important steps in the recognition process [40]. Mentioned below, several categories of iterations that are performed in order to estimate a suitable way for feature selection to distinguish ErrP from noErrP instances, in the EEG data.

a) Each Data Point as an Instance

In this category, single trial ErrP and noErrP signal portion of interest have forty two data points as shown in Figure 22 (a). Each of these data points are considered as independent instances that either belong to ErrPs or noErrPs class without considering the temporal dimension of the signal. For example each ErrP and noErrP signal portion of interest has in total forty two instances belonging to ErrP and noErrP classes respectively. Therefore with this approach, each signal is transformed into forty two instances of one dimensional feature vectors. After the classification result is obtained, a voting scheme is used to have the result for entire test signal, e.g. if more than 50% data points of a signal are classified as ErrPs, then the test signal is classified as ErrP signal else it is recognized as noErrP signal.

b) Dimensionality Reduction of Each Signal

In this method, dimensions or features in signal portion of interest are reduced depending upon the following scheme. For each individual signal, the data points are compared against each other and if they fall in neighbourhood of each other i.e. lie within a pre-set threshold value, then those points are considered as redundant. Redundant points are discarded keeping only one of those points or, in other words they are merged to a single value. Hence the remaining data points of each signal are considered as important features and each of these points are treated as independent instances neglecting the temporal dimension. Similarly this dimensionality reduction scheme is carried out for testing data as well, i.e. it reduces data points which fall in

43

neighbourhood of each other considering all of them as a single point. The threshold is set here as 0.01μv. Similar to the previous category mentioned above, the voting procedure is again carried out here to classify an entire test signal depending upon the classification result of all of its important constituting features.

c) Each Data Point in the Time Series Signal as a Feature

As mentioned earlier, every signal portion of interest contains forty two data points here. Each of these data points is considered as a feature of the signal and therefore each instance of ErrPs and noErrPs has a feature vector of forty two dimensions.

d) Addition of Temporal Feature

It is observed that the time difference between maximum and minimum of average ErrP signals are different from that of average noErrP signals which could be seen in Figure 22 (b). This temporal difference is considered as a feature of the signal and proportional to the difference in index of maximum and minimum value of the respective signal. Therefore, this index difference is estimated for each signal and considered as a feature along with the other forty two features mentioned earlier. Therefore these sums up to a feature vector of forty three dimensions in each instance of both ErrPs and noErrPs.

e) Important Features Only

Among the earlier mentioned forty three features extraction method, a technique is implemented to retain only important features of a signal and use only those selected important features during classification. Classifiers in sequence are developed based on only one feature of all signals followed by another until all the features are used in separate classifiers. Each classifier is validated using the validation data set. If a classifier contributes to at least a pre-set percentage which is set as 30%, for correct classification of both true negative and true positive, then the feature associated with that classifier is considered as biased to correct class feature. Feature might also exist that are biased to opposite class which means feature that helps in classifying the opposite class for both categories of signals instead of predicting the class they belong to. Therefore to detect possibilities of these opposite class biased features, the classifier should have an opposite class classification result above a defined value which is set as 70% for both true negative and true positive signals here. Among total of forty three features including the temporal dimension, those which are biased to correct and opposite classes need to be determined. For the sake of it, a set of forty three linear classifiers are trained separately using only one feature followed by another feature of the signal. All of those classifiers are checked individually against their performances with the validation set. A classifier is considered as poor classifier if it doesn't contribute to either above 30% or above 70% for correct and opposite class classification respectively, for both ErrP and noErrP signals in the validation set. The corresponding feature associated with the poor classifier is regarded as not important and hence is discarded before training the final classifier to be verified with the test data. Important features of average ErrP and noErrP signals are shown in Figure 22 (b).

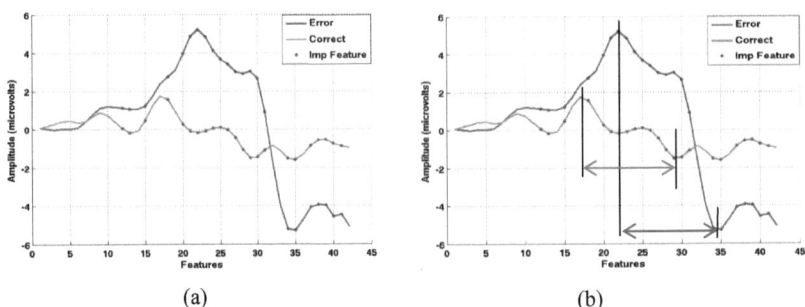

(a)   (b)

Figure 22: ErrP (Error trial) and noErrP (Correct trial) signals (a) Avg ErrP & noErrP with Important Features, and (b) Average ErrP & noErrP with Temporal Difference Feature

3.5 Support Vector Machines Classifier Training

a) Randomization of Data

True negative and true positive data are randomized separately, 10 times, in order to test the result for different training, validation and testing sets. The final test result is calculated as the average outcome of each randomized set of data.

b) Normalization of Data

Each feature needs to be normalized in order to keep the feature values bounded within unity, for the sake of assigning equal weight or importance to all of the features. Also keeping the values within a small range allows the kernel to occupy less memory during kernel function operations which essentially uses calculations like dot product of vectors.

c) Labelling of Data

Each of the available data are labelled as '-1' if it belongs to ErrPs and as '1' if it belongs to noErrPs.

d) Splitting of Data

The ErrP and noErrP data are split into 80, 10 & 10 ratio for training, validation and testing sets respectively.

e) Kernel and Kernel Parameter Selection

Among several varieties of classifiers, Linear classifier tries to linearly separate data while a Radial Basis Function (RBF) kernel or Gaussian kernel maps the limited set of features to infinite number of features in order to obtain a hyperplane separating both classes of data when the data is not linearly separable. By doing so, there are better chances of finding a separating hyper plane than finding it with limited available features. RBF classifier is generally considered having better performance due to this infinite dimension mapping, compared to other classifiers like Polynomial kernels [41]. Among various available techniques like Quadratic Programming (QP), Sequential minimal optimization (SMO) and Least Square (LS), SMO method is selected to find out the separating hyperplane. The SMO algorithm, gives an efficient way of solving the dual problem arising from the derivation of the SVM. It is invented by John Platt in 1998 at Microsoft Research [42]. SMO is widely used for training SVM. The publication of the SMO algorithm in 1998 has generated a lot of excitement in the SVM community, as previously available methods for SVM training were much more complex and required expensive third-party QP solvers [43].

f) Training

During the step of important features selection, only linear classifier is used as it is quicker and also has comparable results with RBF kernel. After selection step of important features, both Linear and RBF classifiers are trained using the features. In general, the performance of classifier is very much dependant on the choice of parameters [44]. In the case of RBF classifier these crucial parameters used in the kernel function are box constraint and sigma value. Initially a coarser and later a finer grid search are generally performed on these parameters. These parameters are chosen depending upon the cross-validation accuracy, after iterating on different values of the parameters. Cross-validation, sometimes called rotation estimation [45], is a technique for assessing how the results of a statistical analysis generalize to an independent data set. It is mainly used when determining the tuning parameters and also the method can estimate the bias of the excess error in prediction [46]. In k-fold cross-validation, the original sample is randomly partitioned into k subsamples. Of the k subsamples, a single sub-sample is retained as the validation data for testing the model, and the remaining (k − 1) subsamples are used as training data. The cross-validation process is then repeated k times or k folds, with each of the k subsamples used exactly once as the validation data. The k results from the folds then are averaged to produce a single estimation. The advantage of this method over repeated random sub-sampling is that all observations are used for both training and validation, and each observation is used for validation exactly once. Classification parameters used during training of SVM classifier are listed in Appendix B.

3.6 Classification Performance of Support Vector Machines

Test instances are recognized using both Linear and RBF classifiers separately and the performances are compared against each other. Test results for each data set generated because of randomization, are produced and the average performance is estimated. Also the standard deviations for both ErrPs and noErrPs classification accuracy are calculated. Mentioned below is the average result of SVM classification for different categories of iterations that is carried out in this work. Among the five different categories used for feature extraction methods mentioned earlier, the first "Each Data Point as an Instance" and the second "Dimensionality Reduction" categories have similar performances with average classification rate around 70% for both ErrPs and noErrPs. "Each Data Point in the time series signal as a Feature" which is third in the category list and the next "Addition of Temporal Feature" both have better performances but average classifications are below 75% for both ErrPs and noErrPs. The result obtained from the last category of iteration "Important Features Only" defined above, is superior to all other categories mentioned. Twenty three features are found to be important after following the procedure mentioned under "Important Features Only" category, among the forty three available features for each signal portion of interest and there is no such feature found which is biased to opposite class. The average classification results of this category are 71% with standard deviation of 12% for ErrPs and 73% with standard deviation of 9% for noErrPs with a linear classifier. As expected, performance seems to increase for RBF classifier with average ErrPs classification of 88% with 8% standard deviation and 75% with 7% standard deviation for noErrPs. The RBF classifier while considering only important features from data is the best performer among other types of feature extraction methods and choice of classifiers mentioned here. The result of SVM classifier and Gaussian classifier for the same data produces ErrPs and noErrPs average classification accuracies as mentioned below in Table 3, for 10 fold cross-validations.

SVM				Gaussian	
Linear		RBF			
ErrPs (%)	noErrPs (%)	ErrPs (%)	noErrPs (%)	ErrPs (%)	noErrPs (%)
71 ± 12	73 ± 9	88 ± 8	75 ± 7	95 ± 5	72 ± 13

Table 3: ErrPs Classification Rate for SVM and Gaussian Classifiers

3.6.1 Timing Performance of Support Vector Machines Classifier

The time consumed during training and validation of the linear classifier is around 70 seconds while that of RBF classifier is approximately 110 seconds, for around 300 instances of ErrPs and noErrPs in total. Both linear and RBF classifier produced test results within 0.1 second for a single test instance.

3.7 Future Work on Support Vector Machines Classification Method

The work could be extended to perform a BCI experiment using either P300 or SSVEP paradigm to record user intention with more number of participants and thereafter training SVM with large data set. A finer grid search could be performed in order to estimate better values of box constraint and sigma which would produce higher recognition rates of ErrPs and noErrPs in the EEG data, when radial basis function (RBF) kernel is used for SVM classification. Advanced feature extraction methods from ErrP and noErrP signals in the EEG data, may lead to even better performance of the SVM classifier.

4 SUMMARY

This is the last chapter of this book and it concludes with the summary of the entire work of part 1 (Adaptive BCI) and part 2 (Classification of Error-related Potentials).

The future generations of user interfaces need to consider uncertainties in user behavior and should be able to adapt as per the need of the user, in order to provide increased user satisfaction while using the interface. This study shows a novel use of machine learning technique, to be specific reinforcement learning method to develop an algorithm for adaptive Brain-Computer Interface system in application to navigational tasks. The learning algorithm used here also takes care of change in behavior of the user and outliers in user behavior. As a future work of this learning scheme, the reward function defined in the algorithm could be enhanced by considering cognitive load of the user. We have shown here that such an adaptive BCI is indeed possible and could provide better embodiment feeling to the user, as if the robot body is an avatar of the user, compared to the conventional fixed policy based BCI systems.

Further this work tackles the detection of single trial error-relate potentials in the EEG data using Support Vector Machines. As expected, it is observed that feature selection plays a decisive role for classification accuracy. SVM is found to work well, in parallel to other classifiers such as Gaussian classifier or Bayesian filter methods and could have higher ErrPs recognition rates for RBF kernel with tuned kernel parameters. The timing performance shows that SVM could be considered for real-time BCI applications. Successful detection of error-related potentials indicates the possibility of increasing the reliability and also improving the information transfer rates of BCI systems.

5 LIST OF FIGURES

6 REFERENCES

[1] Christopher and Dana reeve foundation (One Degree of Separation. Paralysis and Spinal Cord Injury in the United States)

[2] N. Birbaumer, A. Murguialday and L. Cohen, Brain–computer interface in paralysis. Current Opinion in Neurology, 21: 634–638, (2008)

[3] Brain-computer interface Wikipedia Page

[4] H. Alwasiti, I. Aris and A. Jantan, Brain Computer Interface Design and Applications: Challenges and Future. World Applied Sciences Journal 11 (7): 819-825 (2010)

[5] P. Pour, T. Gulrez, O. AlZoubi, G. Gargiulo and R. Calvo, Brain-Computer Interface: Next Generation Thought Controlled Distributed Video Game Development Platform. IEEE Symposium on Computational Intelligence and Games: 251-257 (2008)

[6] B. Obermaier, G. Müller and G. Pfurtscheller, "Virtual Keyboard" Controlled by Spontaneous EEG Activity. IEEE Transactions on Neural Systems and Rehabilitation Engineering, Vol. 11, Issue: 4, 422-426 (2003)

[7] A Leap Forward in Brain-Controlled Computer Cursors (Stanford University Engineering)

[8] F. Beverina, G. Palmas, S. Silvoni, F. Piccione and S. Giove, User adaptive BCIs: SSVEP and P300 based interfaces. PsychNology Journal, Vol. 1, No. 4: 331–354 (2003)

[9] D. Zhu, J. Bieger, G. Molina and R. Aarts, A Survey of Stimulation Methods Used in SSVEP-Based BCIs. Computational Intelligence and Neuroscience (2007)

[10] R. Chapman and H. Bragdon, Evoked responses to numerical and non-numerical visual stimuli while problem solving. Nature, Vol. 203: 1155-1157 (1964)

[11] J. Polich, Updating P300: An integrative theory of P3a and P3b. Clinical Neurophysiology, 118 (10): 2128-2148 (2007)

[12] L. Farwell and E. Donchin, Talking off the top of your head: toward a mental prosthesis utilizing event-related brain potentials. Electroencephalography and Clinical Neurophysiology, 70(6): 510–523 (1988)

[13] J. Wolpaw, N. Birbaumer, W. Heetderks, D. McFarland, P. Peckham, G. Schalk, E. Donchin, L. Quatrano, C. Robinson and T. Vaughan. Brain–Computer Interface Technology: A Review of the First International Meeting. IEEE Transactions on Rehabilitation Engineering, Vol. 8, No. 2 (2000)

[14] B. Blankertz, G. Dornhege, C. Schäfer, R. Krepki, J. Kohlmorgen, K. Müller, V. Kunzmann, F. Losch and G. Curio, Boosting bit rates and error detection for the classification of fast-paced motor commands based on single-trial EEG analysis. IEEE Transactions on Neural Systems and Rehabilitation Engineering, Vol. 11, No. 2 (2003)

[15] J. Bollon, R. Chavarriaga, J. Millan and P. Bessiere, EEG Error-Related Potentials Detection with a Bayesian Filter. IEEE Conference on Neural Engineering, Antalya, 702-705 (2009)

[16] S. Young. Cognitive User Interfaces, IEEE Signal Processing Magazine [128], (2010)

[17] M. Schneider-Hufschmidt, T. Kühme and U. Malinowski, Adaptive User Interfaces: Principles and Practice. (Book - 2007)

[18] K. Gajos and D. Weld, Supple: automatically generating user interfaces. International conference on Intelligent User Interface: 93–100, Funchal (2004)

[19] P. Langley, User Modeling in Adaptive Interfaces. International Conference on User Modeling, Alberta (1999)

[20] A. Wang and Q. Ahmad, CAMF – Context-aware Machine Learning framework for Android. Iasted International Conference on Software Engineering and Applications, Marina Del Rey (2010)

[21] R. Sutton and A. Barto, Reinforcement Learning: An Introduction. MIT Press, Cambridge, MA (1998)

[22] S. Singh, M. Kearns, D. Litman and M. Walker, Reinforcement learning for spoken dialogue systems. NIPS, Denver (1999)

[23] Y. Wang, M. Huber, V. Papudesi and D. Cook, User - Guided Reinforcement Learning for an Intelligent Environment. IEEE International Conference on Intelligent Robots and Systems, Vol. 1: 424-429, Las Vegas (2003)

[24] W. Smart and L. Kaelbling, Effective Reinforcement Learning for Mobile Robots. IEEE International Conference on Robotics and Automation, Vol. 4, 3404-3410, Washington (2002)

[25] J. Liu and R. Hoffmann, Zhiphone: A Phone that Learns Context and User Preferences. CSE567 class project, University of Washington (2005)

[26] J. Bagnell and J. Schneider, Autonomous Helicopter Control Using Reinforcement Learning Policy Search Methods. International Conference on Robotics and Automation, St. Louis (2001)

[27] J. DiGiovanna, B. Mahmoudi, J. Fortes, J. Principe and J. Sanchez, Coadaptive brain-machine interface via reinforcement learning. IEEE Transactions on Biomedical Engineering, Vol. 56, Issue 1: 54-64 (2009)

[28] R. Chavarriaga, P. Ferrez and J. Millán. To err is human: learning from error potentials in brain-computer interfaces. International Conference on Cognitive Neurodynamics, Shanghai (2007)

[29] J. Park, K. Kim and S. Jo, A POMDP Approach to P300 Brain-Computer Interfaces. International conference on Intelligent user interfaces: 1-10, New York (2010)

[30] X. Perrin, R. Chavarriaga, F. Colas, R. Siegwart and J. Millán, Brain-coupled interaction for semi-autonomous navigation of an assistive robot. Robotics and Autonomous Systems, Journal of Robotics and Autonomous Systems archive Vol. 58, Issue 12: 1246-1255 (2010)

[31] N. Eagle and A. Pentland, Reality mining: sensing complex social systems. Journal of Personal and Ubiquitous Computing, Vol. 10, Issue 4: 255-268 (2006)

[32] J. Boger, J. Hoey, P. Poupart, C. Boutilier, G. Fernie and A. Mihailidis, A Planning System Based on Markov Decision Processes to Guide People with Dementia Through Activities of Daily Living. IEEE Transactions on Information Technology in Biomedicine, Vol. 10, No. 2 (2006)

[33] P. Ferrez and J. Millán, You are wrong! - Automatic detection of interaction errors from brain waves. International Joint Conference on Artificial Intelligence, Edinburgh (2005)

[34] P. Ferrez and J. Millán, Error-related EEG potentials generated during simulated brain-computer interaction. IEEE Transactions on Biomedical Engineering, Vol. 55, Issue 3: 923–929 (2008)

[35] S. Bengtsson, Detection and prediction of lane-changes: A study to infer driver intent using support vector machines (A study to infer driver intent using support vector machines). Master of Science Thesis, KTH Stockholm (2012)

[36] L. Auria and R. Moro, Support Vector Machines (SVM) as a Technique for Solvency Analysis. Discussion Papers, German Institute for Economic Research, DIW Berlin Discussion Paper No. 811 (2008)

[37] Andrew Ng, Stanford University Machine Learning Course. CS229 Lecture notes (2012)

[38] Early detection of plant diseases and weeds with Support Vector Machines. DFG Research Training Group (Graduiertenkolleg) 722, International Conference Precision Agriculture, Denver (2010)

[39] G. Schalk, J. Wolpaw, D. McFarland and G. Pfurtscheller, EEG-based communication: presence of an error potential. Clinical Neurophysiology, Vol. 111, Issue 12: 2138-2144 (2000)

[40] X. Wang and K. Paliwal, Feature extraction and dimensionality reduction algorithms and their applications in vowel recognition. The Journal of the Pattern Recognition Society, Vol. 36: 2429 – 2439 (2003)

[41] A. Chan and C. Peng, Wavelets for Sensing Technologies. Artech House Remote Sensing Library: 184 (Book – 2003)

[42] J. Platt, Sequential Minimal Optimization: A Fast Algorithm for Training Support Vector Machines. Microsoft Research, Advances in Kernel Methods – Support Vector Learning (1998)

[43] R. Rifkin, Everything Old is New Again: a Fresh Look at Historical Approaches in Machine Learning. MIT, Doctoral Dissertation, Sloan School of Management, (2002)

[44] O. Chapelle, V. Vapnik, O. Bousquet and S. Mukherjee, Choosing Multiple Parameters for Support Vector Machines. Journal of Machine Learning, Vol. 46: 131–159 (2002)

[45] P. Devijver and J. Kittler, Pattern Recognition: A Statistical Approach. Prentice-Hall, London (Book - 1982)

[46] M. Tsujitani and Y. Tanaka, Cross-Validation, Bootstrap, and Support Vector Machines. Advances in Artificial Neural Systems (2011)

Appendix A: Markov Agent for Training and Testing of the Algorithm

---

Begin

    Define number of actions, trajectory length   // Number of training instances

    Initialize current state, state→action probabilities

    Calculate cumulative state→action probabilities

    while   *training index < trajectory length*   **do**

        Generate random number   // Normalize the random number

        while   *action index < number of actions*   **do**

        if   *random number > cumulative probability (current state → action)*   **then**

            if   *random number = 1*   **then**   *next action ← action index*

            else   *next action ← action index + 1*

            end if

        else

            next action ← action index

        end if

        end while

            trajectory [training index] ← next action

        if   *action = low level*   **then**   *do nothing*

        else   *current state ← action*

        end if

    end while

end

Appendix B: Parameters Set for Support Vector Machines Classification

Karush-Kuhn-Tucker (KKT) violation level	0.01
KKT Tolerance Limit	0.01
Maximum Number of Iterations	20000
Kernel Catch Limit	10000
RBF Sigma	29
RBF Box Constraint	2
Matlab Version (svmtrain)	R2011b
Operating System	Ubuntu 10.10

Printed by Books on Demand GmbH, Norderstedt / Germany